Motel Prius
by Guy Lane
July 2020

If Life Feels Bleak, It's Because Our Civilization is Beginning to Collapse

Umair Haque - 4 July, 2020

The end of human civilization is now easy enough to see, over the next three to five decades. It's made of climate change, mass extinction, ecological collapse, and the economic depressions, financial implosions, political upheavals, pandemics, plagues, floods, fires, and social breakdowns all those will ignite.

oOo

Vita plans to avert this grim future, and instead usher in the Verdant Age, where human civilization and the Living Planet thrive in synergy, deep into the Long Future. To do this, Vita seeks to enrol 53 million people to identify as Vitan.

Are you one of those people?

Preface

Set over 8 weeks beginning in May 2020, *Motel Prius* tells the story of a *Minister of the Vitan Religion* (an environmental scientist called Guy) driving – and living in – a black Toyota Prius hatchback.

Motel Prius is a tale of a personal journey to relocate goods in storage that takes Guy from Brisbane to Sydney to Townsville, and then south to Byron Shire: a crazy, planet-killing trip of over 5,000 kilometres, and hundreds of litres of petrol.

Along the way, Guy has adventures and shares philosophical and spiritual insights in Bayview, Byron Bay, Brisbane, Sunshine Coast, Yeppoon, Townsville, Magnetic Island, Mackay, and Cooroy – and many other places, in-between.

The journey is not described chronologically, and names and details have been changed to protect the identity of the guilty and the shy.

Motel Prius brings together a collection of themes including: Covid-19, the technological wonder of the Toyota Prius, the practicalities of living in the car over an extended period, Vita Religion, abrupt climate change and biosphere collapse, and the looming – *and yet wholly avoidable* – extinction of humans, and most life on Earth.

The story avoids normal tropes of narrative and structure to express the world how it really is, right now: a complex, disjointed jumble of memes. We call this genre: *a Covid Story*.

The biosphere is dying, and *Motel Prius* takes just a few hours to read. Please act accordingly: *Take a look. Be careful.*

Bill Spee
Minister of the Vitan Religion
www.thinkvita.org

Chapters

Cane Toads 1

The Sea Chest 6

Transformation 11

"WTF?" asked *Vitae-planeta* 17

How the Prius Works 24

The Verdant Age 29

The Antisynergy 32

A Squillion dri.bots 36

Sailing to Magnetic Island 41

Vita Philosophy 48

The Prius Wolf 59

Save the Whale 70

Motel Prius Attire 76

Organising Motel Prius 82

Ballina Golf Club 91

Tear Down The Statues 113

Yoga	117
The Boulder	126
The Northern Rivers	132
The Red Cats	137
The Happy Ending	143

Cane Toads

There is a road in Australia adjacent to a pond in which cane toads dwell. The cane toad is an introduced species that has wreaked havoc on Australian wildlife for nearly a hundred years. The toad is harmful because they breed rapidly, attain a large size, and are toxic in all stages of their life cycle: the eggs, the tadpoles and the adult toad are all poisonous.

The cane toad eats just about anything that it can fit in its big mouth; and this includes both native wildlife, and the food that native wildlife eats. Those animals that seek to eat the cane toad quickly find that the glands on the neck of the cane toad exude a milky substance that is fatal to consume.

If there were an ecological scoreboard, it would read: *Australian Wildlife 0 - Cane Toads 1.*

While the cane toad has all the hallmarks of an environmental psychopath, in one way it is beautiful. It has a distinctive call, and if it calls from the water, it sets up a vibrational pattern, ripples on the water surface that emanate away from its body, sending an undulating symmetry of waves that waft outwards and lap against the bank of the pond. There's a video of this on facebook, somewhere. It's quite extraordinary.

The trilling toads can be heard from the road. On that road is what looks like a piece of dried leather. A few feet away lay another piece of leather, and another. These are the desiccated, flattened corpses of cane toads run over by cars – some deliberately, some accidentally.

While it is true that cane toads have wreaked havoc on Australian wildlife, compared to cars, cane toads are amateurs. It's like the difference between a can of fly-spray and a B52 bomber loaded with a thermonuclear weapon. If there were a scoreboard for what the organisms that drive cars have done, it would read: *Australian Wildlife 0 - Cars 1,000,000.*

Despite the distinctive trill, cane toads are widely despised by the humans who live in this country because (ironically) they are foreigners and they kill native wildlife; but also because many Moderns just hate small animals.

Vita regards Western people as broadly having three ways of seeing the world. *Conservatives* want the world the way it was, *Moderns* like the world just the way it is. *Cultural Creatives* strive for a better world. Most people are Moderns, and about a quarter of the Cultural Creatives are Vitans, they just don't know it yet.

There is a meme - *a memory gene* - that has infected a proportion of the

Australian population. It says that cane toads should be killed randomly without any strategic planning that could potentially improve the wellbeing of the ecosystem. In some parts of the country, killing cane toads has become a sport.

For others, cane toads are a given evil that ought to be left alone despite the havoc they cause. A common argument at barbeques is that killing cane toads is wrong because they are living things; even though eating sausages (i.e. formerly alive pigs) at barbecues barely rates a mention.

All this simply shows that humans are able to comfortably hold two opposing beliefs. Without batting an eyelid, for better or for worse.

Pulled-up on the side of the road adjacent to the pond full of cane toads is a black car, a second generation Toyota Prius. Peering from the driver's side window at the flat toads is a Vitan Minister.

You can tell that he's a Vitan Minister because he wears a black cotton t-shirt, blue (or black) jeans, and black shoes. Around his neck is a laser-cut stainless steel pendant in the shape of a Quenn. The Quenn pendant is called a *Quendant*. Further affirming his status, he carries in his pocket a Vita name card with the words: *Minister of Religion*.

The Vitan Minister ponders the flattened toads for a while. He nods approvingly, winds up the electric window, and drives on. He has a mission to fulfil, and the dead toads are a distraction.

The Sea Chest

The woman on the end of the phone had a calm, soothing voice. She told me that her name was Sue.

I told Sue that I needed to drive from Brisbane to Townsville to relocate the contents of a storage shed. This shed was a trove of my personal effects: company accounts, a futon bed, kitchen equipment, diaries, books, CDs, and a Sea Chest full of treasure.

Ahhh! My Sea Chest, full of treasure.

My treasure wasn't gems and jewels as one might expect of a salty pirate, but fabrics (pashmina and silk), statues (a clay horse statue, a bronze, and a faux-jade Chinese dragon, one of a pair). Also in the Sea Chest were diaries, photos, documents and books. Most of the stuff that I owned in the shed was of little value to me, but the chest and its contents was *my precious:* a panoply of artefacts that spoke to my public, private and secret lives.

I had been hankering after that Sea Chest for the whole eight years that it had been in storage. The fate of that wooden box was bound tightly to the narrative of my life, for I am one of those people who never really took root anywhere, and drifted purposefully from one place to the next.

I was born in England in April 1967, and in June of that year I was living in Melbourne, Australia. Since then, I have moved every 6 to 12 months or so, sometimes by choice, other times forced by circumstance.

At age 53, eight years after parking the Sea Chest in the shed, I had an insight that I needed to throw a metaphorical pin at a map, move to that location and set up a base; a base from which I could continue to purposefully drift.

The pin landed in Byron Shire, northern New South Wales, part of a region delightfully named the Northern Rivers. I had friends in the area and was offered a place to store the Sea Chest (and the other boxes and crates) until I found a base. The Sea Chest was in Townsville, North Queensland. At the time, I was in Brisbane, but before I went to North Queensland, I had to visit Bill in Sydney. Like me, Bill was a Vitan Minister.

After visiting Bill, all I had to do was drive from Sydney to Townsville to retrieve the Sea Chest and cart it to Byron Bay. Standing in my way was four and a half thousand kilometres of roads, Covid-19 lockdown, abrupt climate change & ecological collapse. These challenges didn't faze me because I was the master of *Black Beauty*, a Generation II Toyota Prius hatchback. The car would survive anything because it was built in 2008 –

the year the Global Financial Crisis began. Plus, it came to me via an angel (do Vitan's even believe in angels? It's complicated).

Sue, the woman on the phone, told me that the Covid-19 lockdown had been eased, and I was permitted to travel throughout Queensland as long as it was for business purposes. I had a mountain of tax documents in the shed, so this seemed to ethically and legally justify the trip. My plan was to do the ridiculous amount of driving in legs, overnighting on the way and sharing Vita.

In my dilly bag, I had six Quenns. The Quenns were laser-cut stainless steel, adapted into necklaces with a thin stainless steel wire and magnetic clasp. Put together, these formed the Quendant. On my journey, I could share the Quendants, and maybe sell a few to help support the cause.

To ensure I had a safe journey and somewhere to sleep, I converted my Second Generation (Gen II) Toyota Prius hatchback - *Black Beauty* - into a motel.

Motel Prius.

Motel Prius

Transformation

In order to transform a Gen II Prius into Motel Prius, there are two critical steps that need to be executed. The first step is to address the *Spare Wheel Shortcoming*, and the second is to build a comfortable bunk for sleeping in.

It turns out that the Gen II Prius sold in Australia doesn't come with a proper spare wheel. Instead it has a pitiful alternative called a *jockey wheel*.

I discovered this when I hit a kerb, one day, and flattened the front passenger side tyre. When I went to check on the spare, instead of seeing an alloy rim and thick tyre – the analogue of the one I had just ruined – what I saw instead was a feeble, thin tyre on a bright yellow rim.

The tools provided for changing the tyre were woeful, too, and what should have been a quick *'bang and she's up!'* turned into a painfully slow exercise to replace a flat, good tyre with a total donkey.

To rub salt into the wounds, the flat front tyre barely fitted into the tyre well as it was both too wide and too deep. On top of this, the damned jockey wheel was flat!! So, I had to drive a block with a flat, substandard wheel to a service station to get the damned thing pumped up. *Sheesh!* No way was I going to drive thousands of kilometres on Australian country roads with a runt wheel as a spare. I was going to fix the *Spare Wheel Shortcoming* before the big drive began.

So, I visited a wreckers yard to buy a real spare wheel. I found a yard run by some young Indian guys, and there were two Gen II Prius in there. For $50 I got a wheel with decent tread. I was happy, but when I got it home, I remembered the drama with the tyre being too big for the tyre well! *Damn!*

Motel Prius

I got around that by deflating the spare. Deflated, it was possible to squeeze the wheel into the well although this wasn't without problems, as one side of the tyre rubbed against a flange on the battery housing, and the other side of the tyre rubbed against the electrical wiring on the inside of the boot. So, note to self, move the spare in and out of the well with caution. I bought a 12-volt pump that hooked into the cigarette lighter power supply housed inside the centre consul. Problem solved.

The next issue was that under the floor in the back of the car was a plastic storage box that sat over the jockey wheel. Now, with a full thickness wheel, the box rode high, and the floorboards didn't sit flush. So I ditched the box, only to find that it was part of the structure that held the floorboards up.

The solution was to get an electric jigsaw tool and cut a circle out of the storage box that would fit over the wheel. There was plastic fibre soundproofing material on the underside of the box that needed to be ripped away to make way for the jigsaw cut. So, I lost a bit of storage space, but the floorboards were supported. Phew! Fixed.

With the *Spare Wheel Shortcoming* resolved, I was free to add a bunk – somewhere to sleep. The back seat of the Gen II Prius seats three people, and the back seat is split into two pieces with a thin, single seater on the passenger side, and a thick double-seater on the driver side.

I triggered the double seat-back, and leant it forward. This opened up a bunk space that I measured at 74 cm at its narrowest point. It was suitable for my lean frame. I climbed in the back to see if there was enough length room to accommodate me in a prone position. I found a problem.

The top of the back seat, leant forward, didn't reach all the way to the back of the front seat. This meant that there was a gap where my pillow would be. The gap needed to be filled or the pillow would fall into the foot-well behind the driver's seat.

I paid a visit to the local K-Mart and found a plastic inflatable single mattress that was exactly the right width and approximately the right length (a bargain at $7). I had to buy some more plastic (a pump) to inflate the mattress. When I pumped the mattress up in the undercover car park, I found that it was so rigid that it didn't matter that it wasn't supported under the head end (the gap mentioned above).

I had planned to visit an opportunity shop to buy some warm blankets, but the Covid-19 lockdowns had rendered them all closed. I imagined a policeman saying, *"This opportunity shop is dangerous. Go to K-Mart."*

At K-Mart, I bought a fluffy blanket and a soft toy in the shape of a whale to use as a pillow. The blanket was blue, and sort of represented the ocean with the whale floating on top of it.

The first night that I spent sleeping in Motel Prius was an extraordinary experience that I will describe later. It was soft and warm and comfy and awesome. The best way to describe it was the old adage: *snug as a bug in a rug*. And it was on the third night sleeping in Motel Prius that I discovered the wisdom of *Antisynergy*. More on this, later.

It worked! I had converted my Gen II Prius into *Motel Prius*. Unfortunately, there was a serious downside.

Motel Prius

"WTF?" asked *Vitae-planeta*

A few days after I had birthed *Motel Prius*, I had a grim insight into what I had done, and I imagined a situation in which *Vitae-planeta* spoke to me.

Just to be clear, *Vitae-planeta* has no capacity to speak, as *Vitae-planeta* is not human. Instead, *Vitae-planeta* is a living organism that hugs the surface of planet Earth, a living organism comprised of everything that lives on Earth plus the ocean, atmosphere and soil. You (the reader) and I (the author) are bonded not just through this book, but as cells in the body of *Vitae-planeta*.

Some people know of *Vitae-planeta* by the name Gaia. There is a subtle distinction between the two that I wont go into, here. Other names for Vitae-planeta are, *Mother Nature*, and the *Living Planet*.

Like it or not, you are a part of *Vitae-planeta*, like a button is a part of a suit. Or a banana is a part of banana bread.

Had *Vitae-planeta* the capacity to speak (which she hasn't – outside of humans speaking for her) this is what she would have said: *"WTF were you thinking, Guy? There are consequences to every action. What have you done?"*

As a result of having transformed a Gen II Prius into Motel Prius, I had negatively perturbed the ecosphere.

Consider the opportunity cost. I willingly spent AU$75 fiat currency on the transform that could have been spent on something else. Think of all the good one could do in the world with AU$75 (about USD$50). That money could otherwise have gone towards planting a rainforest, teaching a poor girl to read, or getting News Corp out of Australia.

Motel Prius

Next thing, I had to drive about fifty kilometres to and from the wreckers yard, the car parts shop, and K-Mart, and all this driving, powered by petrol, generated a carbon footprint.

I used about 3 litres of petrol to do that trip. It is not widely understood, but burning a litre of petrol produces about 2.7kg of the greenhouse gas carbon di-oxide, also known by the sexy name CO_2.

It sounds counter-intuitive that burning a litre of petrol that weighs less than 1kg would produce 2.7kg of greenhouse gas, but it does. I could explain the math and the chemistry here, but it would be a distraction from the narrative about Motel Prius. If you are curious about this just google it or ask an environmental scientist (like me).

The point is, all that driving around resulted in 2.7 x 3 = 8.1 kg of CO_2 entering the atmosphere. Specifically, the Prius had emitted billions of carbon dioxide molecules from its exhaust pipe. From that day on, these CO2 molecules would drift in the atmosphere and absorb infrared photons wafting off the planet's surface, increasing the heat content of the atmosphere.

By creating the Prius Motel, I had exacerbated the global warming problem that threatens to annihilate most life on Earth in the coming decades.

Creating Motel Prius was a deeply selfish act that had harmed the organism of which I was a part. I reaped all the benefit, but shared the cost. That 8kg of CO2 would mix in the atmosphere, and the hellish global heating would be shared equally by all living things on Earth.

The next thing to consider is the synthetic plastic tyres that would come to grind across forty-five thousand metres of petroleum bitumen; tyres made of petroleum-based plastic.

To make plastic you need to find oil, and for this you need a seismic exploration ship to blast low frequency sound waves through the ocean and deep into the seafloor, annoying the hell out of marine life (I know this because I used to work on these ships).

Then you need a petroleum-fuel powered drilling ship to drill into the seafloor and open a hole in the rocks that contain the oil. Then you need to pump the oil into a ship powered by bunker fuel. Bunker fuel is a grotesque fossil fuel made of all the trash left over from making petrol and diesel from crude oil. The ship will steam the oil across the sea and then pump the crude into an oil refinery where fossil fuels will be used to heat the oil and separate it into its many different molecular compounds. All the crud will settle to the bottom and that will become bitumen to make roads, and at the top will be the lightweight stuff such as petrol and the feed stocks of the plastics industry that you can use in a Gen II Prius, for example.

As the Prius drives across the toxic bitumen, it sheds tiny particles of plastic onto the road. These particles are so small that you can't see them with the naked eye, but you know they are there because of this simple phenomena: you have to periodically replace your tyres because they wear out. All that worn away plastic has to be somewhere: in the environment.

By driving my Prius, I contravened the first system condition of the *Natural Step*: in a sustainable society, nature is not subject to systematically increasing concentrations of substances from the Earth's crust. By driving my Prius, I had also help breach the Planetary Boundaries *climate change* and *novel entities, and* helped to fund a global supply chain run by mentally disturbed sociopaths who extract petroleum oil from deep in the Earth's crust. And by exhausting the products of combustion into the atmosphere and grinding the tyre dust onto the road, I had liberated this evil into the living skin of Planet Earth: *Vitae-planeta*, the super-organism of which I was a single cell.

What on Earth was I thinking?

How the Prius Works

Despite the fact that driving a Gen II Prius is an act of ecocide, the vehicle was, in its day, a revolution in clean technology.

In Australia, the Gen II Prius wasn't the first hybrid on the road. The Honda Insight and the Gen I Prius came before. What really made the Gen II Prius special was the combination of the hybrid technology and the styling, which made the vehicle popular.

The hybrid technology cleverly matched a big traction battery (located under the back seat), a small and efficient petrol engine, an electric generator, an electric motor, and a special gearbox and computer system that distributed kinetic and electrical energy throughout the drive train. The system was called *Hybrid Synergy Drive*. Nice.

Motel Prius

When the Prius is in READY mode, it samples the engine condition every millisecond or so and decides how to power the vehicle. Consider how the Hybrid Synergy Drive manages the energy flows under a number of driving scenarios.

Imagine, for instance, you are sitting in the driver's seat at midnight, parked, but with the ignition on. You are full of tiger prawns, salad, red wine and magic mushrooms, and wondering why there is a *wolf* looking at you through the windscreen. In this instance, the petrol engine and electric motor/generator will be inactive. Thankfully, you are going nowhere.

Or consider, after a refreshing night's sleep, you pull up to a stop sign, and you drive slowly away. The vehicle will deliver electrical energy from the traction battery to the electric motor that will convert it to mechanical energy transferred through the gearbox to the wheels.

As the vehicle picks up speed, the petrol engine will switch on and supplement the wheels with additional force. As the engine will be providing excess energy, some of it will be diverted through the generator, and recharge the battery. If you then put your foot down to overtake another vehicle, the battery will again discharge through the electric motor, adding additional torque to the wheels.

You pick up speed to overtake, pull your foot off the gas, and the petrol engine shuts off, and the vehicle's kinetic energy is converted into electrical energy to charge the traction battery.

All this takes place without the driver having to make any decisions other than how to drive the car. It all happens at a split second, and a digital display on the dashboard shows arrows pointing this way and that to explain what is taking place under the bonnet. It is sufficiently sophisticated to qualify as magic.

In addition, the car has a very low coefficient of drag – it is very streamlined.

The benefit of all this technology is that the Prius uses less fuel than a vehicle with just a petrol engine. The Prius used around half the fuel of its non-hybrid peers, back in the day.

Today, some petrol-only vehicles compete with the Gen II Prius for fuel efficiency, but back in its day, the Prius was a game changer.

The Prius signalled the beginning of the end of petroleum-fuelled motor cars. It wasn't the end itself as the Prius is wholly powered by petrol. But it showed that the human race had seen a light, and began innovating for a future where motor vehicles didn't pollute the city air with toxic pollutants, nor overheat the planet with heat trapping CO2 gas.

The Prius was a segway, a juncture, a fork in the road. It represented an opportunity for the world to embrace a new technology platform and move towards zero carbon transportation, a stepping-stone towards the *Verdant Age*.

The Verdant Age

The Verdant Age is a foundational concept in Vita, and deserves a short chapter of its own. Verdant means green or 'treed' – as in *'a verdant field is a field full of trees'*. Literally, the expression Verdant Age means the age of trees. However, the way that Vita uses the expression is to mean an age in which humans and the living systems of planet Earth operate in synergy. This implies that the planet is better off because of the presence of humans.

This idea is reflected in at least four scientific frames of reference that I will briefly describe, hoping that if you want more information you will google it.

- Astrophysicist Adam Frank, in his book *Light of the Stars*, says that there are likely hundreds of trillions of civilizations in the universe, and that most don't survive. Those that do survive – the *Class 5 Planets* – get by because the civilization develops a synergistic relationship with the biosphere.

- A science paper titled *Gaia 2.0* has a similar theme, suggesting that the biosphere could be augmented by the sagacious input of the human race.

- There is a scientific concept called *Ecological Civilization* that describes humans behaving in a manner that promotes the wellbeing of the global ecosystem.

- Finally, there is a concept called *'stewardship of the whole Earth system'* described in the 2018 paper *Transitions of the Earth System in the Anthropocene*.

Let me describe in a paragraph or two what Vita means when we use the expression, The Verdant Age.

For most of our existence (about 200,000 years), humans have lived in balance with nature, in the same manner as all the other species. However, since about 70,000 years ago we have deviated from this path. Through our great numbers and our technology, we now dominate the climate and environment in a negative manner – this age is referred to as the Anthropocene.

If human civilization can wake up in time, it is possible that we can use the global economy and our technology in a different way - a way that nurtures the Living Planet back to health. If we can do this, then there is a good chance that we humans can live on this planet for hundreds of millions of years. This is the Verdant Age, and bringing it about is the primary mission of the Vitan Religion.

The Antisynergy

The first night sleeping in Motel Prius was pure bliss, although getting into the bunk was a major challenge. At bedtime, I walked in the light rain from the shed that my friends had converted into a house.

I stepped into the driver's seat, placed my foot on the brake, slid in the electric key and pressed the 'power' button. Motel Prius came to life. The lime green illumination of instrumentation lit-up the cabin with a dull, verdant glow. The engine engaged for a few seconds, then shut off. I fussed around in the front seat for a while, checking that miscellaneous possessions were in their allocated locations. I slid off my boots, leaving them in the foot well of the driver's seat, and prepared myself for bed.

Despite it being cold and windy outside, Motel Prius was warm inside,

Motel Prius

so I didn't feel the need to keep the Hybrid Synergy Drive active to run the air-conditioning system. I turned on the overhead light, and powered down the synergy drive. Now, bathed in a dull orange light, I contemplated the next move: how to get into the bunk from the front seat. I concluded that this was not something that could be resolved by thinking alone, so I got into action moving my body from the front seat to the back where my bunk was located. It was possible, but not easy.

The first thing was to raise and twist my body around so that my right foot could find a place to land on the front passenger seat. This movement tested my dexterity and was not conducted without some muscles being stretched beyond their happy place. Next was to find a firm handhold to take my weight, and then half crawl, half fall into the back of the car. Then, wriggling around like a lizard in a tin, I squirmed further towards the back of the car, with my head pushed against the glass hatch. I was not planning to sleep with my head at the exhaust pipe end, so I needed to rotate a full 180 degrees in that confined space. Possible yes. Easy no.

Eventually, after an exhausting effort, I found myself laying on my back, looking up at the ceiling of Motel Prius, and feeling a most amazing sense of warmth and comfort. The $7 air mattress and the petroleum plastic blanket moulded around me like a cocoon and I let out a long blissful sigh. Outside, the wind blew and the rain fell, pattering on the metal roof a few inches from my nose. I glanced in the direction of my feet to see the glass hatch, and the decorative styling inside the car. I felt like I was floating in a tin can, far above the world, and safe from the harsh external conditions. It was blissful.

Motel Prius

The next night I performed the same ritual with similar results: a good night's sleep. By the third night, the air mattress had let go a bit of air, so I pumped it up, nice and rigid. Too rigid, it turned out, because when it was fully inflated, the foot end of the mattress protruded a few centimetres out the back of the car. So, when I slammed the hatch shut, the metal faces pinched the mattress and all the air leaked out. I didn't know this until bedtime, and when I clambered into the back of Motel Prius on the third night, I was met with a hard surface, and I discovered the ANTISYNERGY: *an air mattress without air is not a mattress.*

guylane.com

A Squillion dri.bots

Heading south toward Sydney from Lennox Head the day after the third night, I ventured into the towns of Yamba, Coffs Harbour and Port Macquarie. Port Macquarie will always remain as scar tissue in my mind because of the dri.bots.

On the way south from Coffs Harbour, I got this idea in my head that I ought to have some of those wet-wipe things, so that I could do mini-cleansing routines as I drove. Like on an airliner where the cabin crew come around with the warm, moist towels and you feel totally refreshed after you rub down your face, arms and neck with them.

I remember some highly organised women I had met kept these things close to hand: excellent for tidying-up after wine and cheese. Wet-wipes were also big in the news in Australia at that time, because members of the public had mistakenly believed that Novel Coronavirus 2 would impact the frequency with which they needed to visit the toilet, and so there had been a run on toilet paper. I saw this with my own eyes: supermarket shelves stripped bare.

As a result of this odd social phenomenon, many good people ran out of toilet paper, and in their desperation, they took to using wet-wipes instead. While wet-wipes perform a similar function to toilet paper (i.e. cleaning things), they give very different results once in the sewer system. One post on facebook showed a plumber dressed like a rapper, drenched in gold chain. You see, wet-wipes don't break apart easily, like toilet paper does, and they clog sewer pipes.

Wet-wipes + Covid-19 = rich plumbers.

Anyway, all that aside, I did a quick tour of K-Mart seeking a mattress that couldn't deflate. I get a bit twitchy in supermarkets, so I made my rounds as quickly as I could.

I found a yoga mat, and I thought, *well if it's made for yoga, it must be sustainable*. As I was approaching the checkout with my supposedly sustainable yoga mat, I saw a shelf filled with boxes of wet-wipes for the astounding price of just $9 a box. This particular brand of wet-wipes was marketed toward babies, and the name on the box was: dri.bots.

Wet wipes, dri.bots, who was I to be choosy? I grabbed the box of dri.bots, paid-up, and departed the store, feeling like I had scored the deal of the decade.

When I got back to the car and opened the box, I realised just how many damned dri.bots I had acquired. There were 8 bags stuffed with like 80 of the damned things. What's that, a million, or something? (Bill calculates 840 at 1.4 cents each. Thanks Bill. It's good to have a Bill on the other end of the phone). There are like thousands of them. And they are everywhere, like cane toads. I go to check the spare tyre and there's a pack of dri.bots in the way. I open the glove compartment and a bag of damned dri.bots falls out. I put my hand in the pocket of my jacket and there's a desiccated dri.bot there.

To make matters worse, I subsequently found that the reason that wet-wipes and dri.bots clog sewer pipes is because they are made of finely woven plastic, not paper. On top of all this, I then find that the damned yoga mat isn't made of a natural fibre, ethically sourced. I had purchased 600grams of polyurethane! More plastic!!

Now, if I had had an ounce of common sense I would have said, "*No, Guy, don't buy a whole box of dri.bots, just buy a single pack. Or, hey, maybe I check the ingredients before I buy the yoga mat."* However, absent that sensible thought process, I committed more of my limited supply of money to the petrochemical industry that is so royally poisoning our planet.

When I arrived in Sydney, I offloaded a pack of dri.bots to Bill, and in return he cursed me with a tin of powdered food, which, like the remaining squillion bags of dri.bots remains in the back of my car, unopened.

Sailing to Magnetic Island

Thousands of kilometres north of Sydney is Townsville where my Sea Chest resided. I couldn't access the storage shed over the weekend, so I accepted an invitation from my friend Dan to sail his 24-foot Seawind catamaran across Cleveland Bay to spend a few days on Magnetic Island.

The vessel has two slender fibreglass hulls, a flat plywood deck and a black fabric matting at the front called the trampoline. Dan is a master yachtsman and I honoured his space as he fussed around, preparing the vessel for the journey. He cast off the mooring lines, motored down the creek, and set the mainsail and jib, and motor-sailed into the shipping channel.

I found a place to witness the master at his work, parking myself in the companionway, the hatch that led down into the starboard hull. The starboard hull was the nerve centre of the vessel. Here was the radio, the stove, the master's narrow bunk (a big bunk for a boat this size, I was informed), the esky (fridge), food, wine and tobacco pouch.

We exited the shipping channel and the island. Finally, the moment I had been waiting for: Dan shut off the outboard engine and the droning noise of the four stroke fell silent, and all that could be heard was *Yachtsound*.

Yachtsound is the noise made by waves lapping against the two hulls, the wind moving past my ears and the miscellaneous noises of a boat under sail. Yachtsound is sacred, and *Yachtsound Bliss* is a name I give to the time between the engine shutting off and someone ruining the tranquillity by opening their mouths, and talking.

When I am the skipper, I am disciplined in maintaining Yachtsound Bliss. On my old boat I sometimes managed to get a full five minutes before one of the passengers started talking started, generally saying something like, *"Oh, isn't it lovely quiet without the motor on."*

I closed my eyes and let my body rock gently in time with the waves, enjoying a form of *Vitan Meditation*. It didn't last long before Dan shattered the tranquillity. Despite being an intelligent empath, the skipper had a few thorny edges.

"Oy!" he demanded, waking me from my bliss. He was standing there, looking as though he had been interrupted free passage down below. He said, *"There's only three types of people who sit in the companionway: Admirals, boat owners and assholes; are you an Admiral or is this your boat?"*

Yachtsound Bliss lasted about forty seconds on that trip; neither a record for length or brevity.

We moored at Magnetic Island and Dan went about his normal business instructing novice yachtsmen about the proper way to conduct themselves aboard a vessel.

Later, we wandered ashore looking like yachties: unkempt hair, barefoot, pants rolled to the knees. Dan wore a Sea Shepherd hoodie, and I wore an RM Williams waistcoat that I had bought in Tamworth from an opportunity shop for $15 during an adventure from another time. We crossed the bitumen car park to the supermarket where we purchased 300 grams of petroleum-based plastic with food wrapped inside. Dinner consisted of wine, roll-your-own-tobacco and things to eat.

As night came, the air cooled considerably and I put on a grey woollen jumper that I had bought from an opportunity shop in Frinton-on-Sea, Essex, England, a few years before.

Motel Prius

Earlier in the day, Dan had chastised himself for leaving the blankets in the back of the car. A few nights before was the coldest May night on the historical record in the region. This was an event scarred into the minds of North Queenslanders who are used to being bathed in warmth. So he bought me a blanket from a shop. Anyway, he redeemed himself by buying a blanket on the island.

I was fine, tucked up in the after-bunk on the port side of the Seawind, surrounded by fibreglass; sort of like a poor-man's Motel Prius. I slept in my clothes, waking periodically through the night to adjust my position. I had that sour taste in my mouth that comes from smoking cigarettes made by corporations, and the mild numbness that comes from consuming more wine than the health regulations advise.

I don't normally smoke, but it is something of a cultural tradition for me when I am hanging with Dan. I have learned how to turn tobacco on and off. Not so much the wine drinking. That's parked in the on-position, for the time being, at least.

Before bed, the night had been punctuated by the rowdy noise of six Moderns partying nearby. Dan was Vitan, and I had at one time considered sharing Vita to the other humans who I met on that sailing trip, but they all seemed pretty happy with the world as it was.

This is the tell-tale sign of a Modern, and I generally don't waste time sharing Gaia Theory and *Vitae-planeta* with Moderns. They just don't get it. Yet.

They will, one day, maybe. One day after some freak weather event burns down or washes away their most cherished possessions, or their family. I'll be ready to share then, and, all going to plan, I'll have a sophisticated organisation and all the communication materials that can help them to have an Ecophany – an ecological epiphany – in short order. In Vita-speak, this is called *Meeting Your Red Cat*. More on this, later.

Vita Philosophy

When the moment comes that someone takes and interest in Vita, and truly starts to listen, this is one of the important things that I would tell them.

A new era of freak weather and climate extremes is upon us, and just like with Covid-19, our societies are unprepared. It is one thing to know about pandemics, and quite another to experience one and manage it.

Since the mid-1970s, there have been multiple warnings about climate & ecological collapse from scientists and activists; and yet the mainstream public, corporations and governments have stubbornly refused to act. So, just like Covid-19 slamming into Wuhan Province, Northern Italy and New York City (to name just a few places), the whole world is going to get smashed by climate & ecological collapse, completely unprepared.

Covid-19 is a good wake up, but what's coming down the pipeline will make the pandemic look like a nice day on the beach.

To adopt an appropriate philosophical approach to the human and ecological condition in this phase of the *Anthropocene Epoch*, Vita suggests that there are two frames of reference that we need to be comfortable with.

One frame of reference is that we are on the downward slope of a massive extinction event called the *Sixth Extinction*, due to climate change and the destruction of ecosystems. We may have already triggered *the cascade of climate tipping points* that drives global temperatures into the Hothouse phase, in the coming decades. If you want evidence that this is already happening, consider the contemporary forest fires in Australia, the sea ice retreating and methane belching out of the Arctic (*it reached 38 degrees Celsius in the Artic Circle during the time I was writing this book*), and the carbon sinks failing, for example the Amazon Jungle and the phytoplankton carbon pumps in the oceans.

We may simply have already condemned most life on Earth to extinction, and this includes our race – *Homo sapiens* – being amongst the first to go, along with all of the other mammals.

We may simply have no capacity to fix the mess. Imagine that we all woke up at the very last moment and got to work *euthanizing the fossil fuel industry*, and regrowing ecosystems to suck three trillion tons of carbon out of the atmosphere in a few decades. Imagine that we were able to reinvent politics and economics in time, but the physics of the Earth system simply didn't permit it.

Motel Prius

We may simply have already condemned most life on Earth to extinction – a rerun of the Permian Extinction, over the coming decades.

So, you ought to be able to sit with that. Find your peace with it. Find a way to let go that this is meaningful. If you can find peace having adopted that as a reasonable frame of reference for the future, then you're empowered to go about your Vitamission without fretting about whether it's going to work, or whether it's going to make a difference.

The other philosophical position calls upon you to accept that *today is better than tomorrow*, and that while global ecosystem conditions are degrading rapidly and humanity is going to a very dark place, we might actually be able to turn things around.

Let's not kid ourselves about the nature of the task. There is a massive amount of work required just to stabilize the patient: the Living Planet. To actually bring about the Verdant Age is a mission that can only be understood as *making patterns out of chaos*. And yet, it may be possible.

So, Vitans need to be comfortable with two philosophical positions:

(1) we will inevitably destroy our race and most of nature, and;

(2) we are going into a place of madness and chaos, but our race will survive in the end.

I don't think that there is any benefit in having a philosophical position that says that there is no problem. If you hold that position, then I suggest then you're not paying attention to *Vitae-planeta*. You don't have your feelers out for the well being of the host. You're paying attention to something, but it's not the Living Planet.

So, to recap: the two positions that Vitans need to be comfortable with are that we may have already condemned life on Earth to extinction – a rerun of the Permian extinction over the coming decades.

The second position is that things are going to degrade for decades, but at some point we will turn things around and start regrowing nature.

These two positions need to be explored, probed with scientific discipline, contemplated, discussed, shared. These are the two pillars of *Vitan Philosophy*.

And you really should get comfortable with these two prospects because you're not really in control of much. When I say *you*, I mean we. We people, individually, don't really have much capacity to change global events.

Sure, one person can do amazing things. Think of the sagacious Greta

Thunberg, or Mahatma Gandhi, Martin King, James Lovelock, and all the leaders who created awareness amongst millions of people.

But all those leaders combined haven't stopped the biosphere from dying.

So there's a certain point at which you realize that it's sort-of immaterial what you do. You, personally, don't get to change the outcome.

You could stay at home playing tiddlywinks and drinking green tea. Or you could lead a rebellion against the sociopaths. In a way, it's all the same thing.

With that said, there are global movements: ideas that spread around the world, sparking action. It's when thousands and millions of people get aboard an idea that the world changes.

What's clear over the last couple of weeks is that beige-faced police murdering people of darker colour is common, and up until now, everyone just shrugged and said, *"That's how it is"*.

Then all of a sudden there's a picture of a dark-skinned guy getting his neck squashed by a policeman with his hands in his pockets, as though there was absolutely no ethical, moral, or career consequence for cold-blooded murder.

And that image is all that is needed for everyone to say: *"This is crazy. This is not how it should be. There's a better way. Let's take risk and action to make it better."*

And it's more than police murdering people who don't have pale, beige skin colour. It's also about the statues that honour genocidal psychopaths. Here's an idea, let's get rid of the statues of the slave traders. Let's get rid of the statues of the people who facilitated mass child-rape. Lets get rid of the statues of people who foster Earth-murder. It's actually quite easy to do. Just get a few brave folk together with enough righteous anger and some rope, and keep pulling until either the statue comes down, or the police arrive.

Now, people have been trying to spread these ideas for decades, without much success. And then all of a sudden: *Bang!* It takes off. There's an inciting moment: a picture of a man being murdered by a cop with his hands in his pockets. Then the idea that many people have long been thinking turns from a gas to a solid, and direct action follows.

And yet - meanwhile - the humans continue to murder the Living Planet. We need a movement to fix that, a sustained movement: a movement that lasts a million years.

We need a movement where people start seeing themselves as part of the whole, instead of individuals in an environment.

Motel Prius

Black lives matter because life matters.

But black lives matter particularly, because at this moment in time, they can't breathe. We need to support this movement, and then expand on it, because on this planet, *life is finding it hard to breathe*.

We need to spark a *Million Year Movement for Life*. MYM4L.

But you can't choose to make that movement happen. Black Lives have mattered for all time, but only recently did cities burn to demonstrate that.

You can't deliberately spark a revolution – revolutions are fickle. But you can help fuel them. The fuel comes before the spark.

Everyone wants to spark a revolution, but they don't have the influence and power. What we can do is to throw the seeds around. Soften the ground. Stockpile the fuel. So when the inciting spark happens, it sparks a wildfire.

So, for Vitans who want to bring about a planet where humans aren't destroying the biosphere, just keep putting out the messages, sharing ideas with people, get better with marketing and communications, find new resources, keep putting it out on social media. Talk about Vita, share Vita, and then one day there will be an inciting event and it will take off overnight. And we will be on our way towards the Verdant Age, in no time.

The Prius Wolf

It's not possible to say whether or not I ever began micro-dosing magic mushrooms, but assuming that I did, consider what might have happened on the first night. The following is spoken in the present tense and first person, to be consistent with the rest of the book.

I took the first capsule at about 6pm, and then enjoyed a full night of wine and good company. The capsule contained a mid-level micro-dose of magic mushrooms that contained an active ingredient called *psilocybin*. My 'treatment' regime was to take one of these for two days, then take a day off. This two on, one off regime was to continue for a month. 20 capsules over 30 days. After this, I would go to a 50% higher level of magic, but keep the same 2/1 pattern. According to my carefully crafted shroom-strategy there would be no more to be gained from the lovely mushrooms at that point; and there would be no point to continue taking them… Until there was.

After my night of wine and good company, I returned to Motel Prius. This was my first ever experience with psilocybin, and seated in the driver's seat of Motel Prius, I was contemplative. I activated the music system and searched my petroleum-plastic folder full of petroleum-plastic CD discs. The one I wanted was by Bjork. In particular, I wanted to listen to her song *Bachelorette* in which she opines: *"If you forget my name, you will go astray, like a killer whale trapped in a bay."*

I played this song over and over, and at some point realised that the Prius instrumentation lights were too bright. I had heard that the Gen II Prius came with a switch to dim them, but I was in no mood to go searching online or through the hard-copy manual in the glove box to figure it out.

Maybe this is the point at which the mushrooms kicked in. Faced with this dilemma (too much dash light and no mojo to read the manual) I struck upon a cunning plan. Under my left elbow was the lid of the centre consul, and under this was a bag of dri.bots. I whipped that bag of dri.bots out and pushed it against the middle of the instrumentation panel.

The glaring symbol, telling me, pointlessly, that I was travelling at *0 km per hour* was eliminated from my view. I observed that there remained a number of small light sources emanating from the dashboard of Motel Prius.

Amongst those lights was a clock informing me with lime-green illumination that it was 11:23, and another indicating that the Prius was in 'Park' mode. Both these sources of visible photons were at the 9 o'clock to my position: just to the left.

The dri.bots had done a good job of squashing the glare of the instrument lights, but not a perfect job. Two light-sources remained, one on either side of the dri.bots bag.

To the left of the dri.bots bag was a word in bright green full-caps saying 'READY'. And to the right of the dri.bots was a red symbol indicating that the park brake was activated. The park brake symbol was an exclamation mark inside a circle, and outside the circle, on either side, were two quarter-circles.

The READY and the PARK BRAKE symbol looked like the eyes of an animal! The light from the eyes reflected on the dashboard, and this reflected off the windshield glass in a hazy fashion. The combination of the 'eyes' and the reflection of the reflection gave the impression of the eyes and ears of a wolf.

It wasn't subtle.

I sat upright in surprise and exclaimed: *"WTF? There's a wolf looking at me."*

Now, this was not a scary realisation; quite the opposite, in fact. The wolf was a welcomed guest. Not just a guest; the wolf was watching over me. Guarding me.

Guardian Prius Wolf.

Nothing I could do would upset the wolf because what I had uncovered was *the wolf inside me*. The wolf was of me. The wolf was me. I was the wolf. I am part wolf. I never knew that before the shrooms.

"Hello Wolf," I said. And the wolf said nothing in return. I wasn't upset by the silence: it kind of made sense. It's like talking to yourself; you don't need to reply.

I have tried hard to explain what micro-dosing magic mushrooms has done for me. Most people remember the wolf story because it fits a preconceived idea that magic mushrooms make you hallucinate. Do they? I don't know. All I know is that on the first night that I started micro-dosing magic mushrooms, I realised that I had a wolf looking after me. If you want to know how powerful that is, watch a movie with wolves in it. There's one called *The Grey* with Liam Neeson that sums it up. You want to have wolves on your side.

Another way of understanding MMM is it's like ingesting a key that drifts around your psyche opening boxes of ideas that were previously locked away. Now, with the lids open, those ideas are free to drift around. It also helps to find critical pathways from where you are to where you want to be. And it makes innovation easy. It's like having two of me inside my head to kick ideas around. Things get thought-through in a more thorough fashion.

It also brings a sort of contemplative melancholy; waves of realisation.

There is a 2011 movie called *Margin Call* featuring Zachary Quinto who plays *Peter Sullivan*, a risk analyst in a big Wall Street financial firm. On the day that the axe comes down, Sullivan's boss *Eric Dale* (Stanley Tucci) is let go from the firm. As *Dale* is hustled into the elevator by security, he hands *Sullivan* a memory stick, saying, *"Hey, look. I was working on something but they wouldn't let me finish it. Take a look at it. Be careful!"*

"Be careful!"

Afterhours, *Sullivan* takes the memory stick and delves into the new mathematical risk formula that *Dale* had been working on. After several hours, he puts in some missing data and the formula delivers its result. In short, the market will imminently collapse: a *Global Financial Crisis*.

There is a moment in the film where Sullivan has his realisation. He sits back from the screen and pulls the headphone from his ear. His jaw drops a fraction. These are the physical aspects of realisation. Meanwhile, something else is going on inside his head.

At that moment, neurons that haven't previously communicate with each other, shake hands and say *'Hi'*. *Sullivan* is an empath, so he understands not just that the *'math doesn't add up, anymore'*; he also realises the social impacts of the impending crisis. A lot of good people are going to get hurt.

It's that moment of realisation that has defined my shroom experience: a multiple of *'Ah-ha'* moments. They come and go in small waves, and at unpredictable times.

What is interesting about these realisations is that they are not specific. They are general. I feel as if my broad understanding of the world has risen on these waves. It's like gaining a few extra points of IQ just for holding your mouth in the right position. It's free.

Not entirely free. Everyone needs to get paid, so there is a small monetary cost. But consider how many thousands of dollars we piss away into the wind on all the *muda* – all the futility, uselessness, wastefulness of the dollars we spend. All the junk we buy and then throw into the landfill. Think of the opportunity cost – what do we forgo by spending money on that crap?

There is an English proverb that goes something like this: *But for a nail, a shoe was lost. But for a shoe, a horse was lost. But for a horse, a knight was lost. But for a knight, a battle was lost. But for a battle, a war was lost. But for a war, a kingdom was lost.*

But for a nail a kingdom was lost. A kingdom for a nail.

Everyday, we needlessly throw nails away, and *we are losing the Kingdom of Life*, as a result. We Western humans are among the most ecologically destructive organisms ever to draw breath on this planet, and as a result the entire living skin of the planet is dying. We are killing our life support system through our consumption of petroleum-based products, and the purveyors of these products have so corrupted our political systems, that the public is offered little alternative, besides abstinence.

One day - hopefully sooner rather than later - you (dear reader) will understand what is coming to us all due to climate and ecosystem collapse, and breaking through the Planetary

Motel Prius

Boundaries. Reflect upon the two principles of *Vitan Philosophy*.

I hope that this book – *Motel Prius* – has the same effect on you as the formulae on *Eric Dale's* memory stick had on *Peter Sullivan*. I hope that you will be able to look at all that is written here, and add in the bits of the equation that are missing for you, so that you can have your own personal "*Ah-ha*" moment.

The biosphere is dying. Act accordingly. That's as simply as I can say it.

My advice to you is the same as *Eric Dale* gave to *Peter Sullivan*.

"*Have a look. Be careful.*"

Save the Whale

My cry for attention about the demise of our race and our planet is not new. People have been talking about this since the mid-1970s, when the science became clear. That's half a century lost to inactivity.

Think for a moment about the early days of Greenpeace. There's this movement, and people are going around and saying that we ought not to murder whales with exploding harpoons. The idea is simply called: *Save the Whale*.

And the regular people who hear about this are like, "*Okay, so what's a whale and why should we save them?*"

And Greenpeace folk say, "*Well, it's like a big fish but it's a mammal, so it gives birth to live young, and breast-feeds the babies on milk. And the babies weight half a ton. These are big animals, right? And people in Russia and Japan, Norway and other places are murdering the whales to make money. And they use a Svend Foyn gun which combines a grenade with a steel harpoon that blasts big, bloody holes in the whales, and they die an agonising, brutal death. And it's wrong because the whales are beautiful. They are sentient creatures. They feel pain and emotional distress, like humans. And they have an important ecological role in the environment as they help regulate the climate by priming the ocean carbon pump. And they have great spiritual and cultural value as denizens of the ocean, and totems. And they are beautiful and majestic, and some of them sing these long, melancholic songs that can go on for days. And they are worth more alive through ecotourism than dead on a plate. And it's just wrong to murder them. And many types of whale have been so brutalised by whaling and other environmental stresses that they are endangered to extinction. And the whale murder machine supports a capitalistic marketplace that benefits the super-greedy psychopaths that run the world so badly. And killing whales introduces bio-accumulated mercury into the human food supply. So, there's a whole bunch of reasons why you should not kill the damn whales, okay?"*

And the guy goes, "Yeah, I get it. Whales are cool, man. We should just let them chill in the ocean and leave them alone."

And the Greenpeace guy says, *"Exactly. So how about you throw some cash into this campaign, or wear the badge, or come and jump on the boat. Do something to Save the Whale, right?"*

In this analogy, the pitch about the need to preserve whales is followed by a request to perform *a particular action*. If I remember correctly, some personal development people refer to this as *enrolment* and *registration*.

Motel Prius

Save the Whale is a rather transactional equation: I give you an insight, and you do something in return.

The alternative to this is the *Vita Approach*. I invited you to consider adopting the Living Planet - *Vitae-planeta* - as the locus of your *spirituality*. And in return I ask nothing of you.

Sure, you might buy a Quendant to support the cause, but I am not telling you about *Vitae-planeta* as a pathway to sell Quenns. I am telling you about *Vitae-planeta* because she is dying, and you have a spiritual duty to defend her.

My helping to relocate your spiritual centre comes with no specific request.

It's not my job to tell you how your nature-centred spirituality ought to play out in your life. Instead, I am just going to leave it with you. Let you sit with it.

And if I perform my Vita Mission properly, you won't get distracted and forget the insight. You will mull all this over, and find your peace with it. And it might be that you never actually do anything as a result of that. And maybe that's just as good as it gets with you.

Or it might be that after a certain period of contemplation and reflection, you undertake a complete upheaval of your life: lead the Extinction Rebellion, take down an oligarch, become a Vitan Minister, or some such thing.

Vita does not demand people to take any particular action. Instead, it seeks to incite people to *align their spirituality with nature*.

And, to be clear, Vita *does not* ask you to ditch your spiritual connection to crystals or christianity, yoga or god; but we do ask that you anchor your spirituality to nature – put it first; and seek to understand nature, not just emotionally, but with the rational part of your mind, as well.

If it is true that having a strong spiritual connection to nature fosters you to take right action to protect her, *then simply by sharing Vita, we help advance the Verdant Age.*

Motel Prius Attire

Day in, day out, I wear the attire of a Vitan Minister, and this has many benefits, particularly when one is living in Motel Prius. The official attire of a Vitan Minister is both simple and repetitive.

Now, just to be clear, at time of writing, there are only two Vitan Ministers, and both are male. To date, there has been no consideration of what Vitan attire for female ministers might comprise. Indeed, there has been no consideration of what the attire for four of the five genders might include. Vita is a new religion, and much is yet to be written.

Those readers who are perplexed by the last sentence might consider that Vita holds a view on gender that is advised by that of First Nations cultures. Vita's position on this thorny issue is still being resolved, but the gist is that there are five genders, broadly describes as:

- Females
- Males
- Females with male characteristics
- Males with female characteristics
- Those people for whom Female/Male isn't a helpful frame of reference.

Vita holds that the non-binary genders are imbued with sacred values of insight that Males and Females can only ever hope to emulate.

In the meantime, the male Vitan Ministers get around in obligatory socks and jocks, and four other uniform elements: an unbranded black cotton t-shirt, blue or black jeans, black shoes, and a Quendant (Quenn pendant). Waistcoats, jumpers and overcoats are optional extras.

Some people have suggested that Vita Religion ought to include some official headwear. This idea is under serious consideration.

In Motel Prius, I carry five shirts, four pairs of jeans (overkill – only really need two or three at most) and socks and jocks to last ten days. You only need one Quendant, although having a spare is a good idea, in case the original is lost, or you inadvertently give it away at cocktail party.

In cool weather where perspiring is minimised, you can reasonably get two days out of the t-shirt and four out of the jeans without becoming socially unacceptable. The idea is to have enough clothes on hand to last you at least a week before you need to visit a laundry.

Just as an aside, there is a troubling paradox associated with the regular washing of clothes. It is socially expected that people wear clean clothes as these beget hygiene and health. However, washing clothes wears them out, dumps nutrients into waterways, consumes electricity and water, and can release petroleum-plastic microfibers into waterways.

The black t-shirts are heavy knit cotton that don't need ironing as long as you fold them neatly, and don't tread or sit on them, or accidentally run over them with Motel Prius.

In short, the Vitan Minister's outfit is neat and tidy. If you keep your hair looking respectable (slicked back with Vitan Hair oil maybe, or just washed daily with warm water), you can pretty much go anywhere in this uniform. Try not to use multinational shampoo, because it contains a lot of really nasty toxic stuff including phthalates.

Try it out. Become a Vitan Minister, dress accordingly, and casually wander into the lobby of a luxury resort and politely inform the person behind the reception desk that you'd like to speak to the manager. They'll pass on the message. The manager will appear, glance you up and down, noticing your smart, disciplined and unbranded attire, and immediately assume that you are moneyed, and thus worthy of speaking to.

Now, if for any reason, the resort manager happens to get a glimpse of Motel Prius, this will only confirm their suspicion that you are one of those reclusive billionaire types who keeps a low profile in order to blend in with normal people.

But! Beware of letting the resort manager see the bunk in the back of Motel Prius, lest they twig that you are sleeping in the car. It would be hard to come back from that, as resort managers can be very cagey, and unable to resolve the contradiction of a billionaire sleeping in a car.

Imagine that!

Motel Prius

With the Vitan Minister's uniform, you can go anywhere. You can hang out with hippies, playing the role of the square dude who's not that square because he wears jeans and talks about Gaia Theory.

You can go to art gallery openings because you actually look like an artist.

You'll get invited to cocktail parties because you blend in as the guy who is so cool that he doesn't need to dress up. Plus, you are wearing a Quendant, so everyone is going to notice that, and they'll ask you what it is. And if you are savvy, you might even sell a few to help fund the mission, but more likely you'll drink to many cocktails and end up giving them away.

Organising Motel Prius

If you want to become a Vitan Minister, and go about sharing Vita, traveling from place to place in Motel Prius, you'll need to be organised. Motel Prius is a car, not a storage shed, and space is limited.

If you are not organised, you'll spend hours scratching around like a bush turkey digging in the dirt, trying to find things. Besides being an inefficient use of your time, this will also annoy the hell out of you and usher on the moment that you yell: *"What the hell am I doing living in a car?!"*

Over my weeks living in Motel Prius, I have developed a simple plan for how to use the various parts of the car for storing the stuff I need.

Front Driver's Seat

This is the nerve centre of Motel Prius – let's call it the Reception Desk, or just

'Reception'. From here, you drive the car, and here also is the computer workstation (where I am, at this very moment, writing these words on my laptop resting between my lap and the steering wheel). From Reception, I can activate the CD player (to listen to *Bjork* sing about killer whales, for example), and activate the air-conditioning system. Here, also, I can, on occasion, observe *Prius Wolf*.

The foot-well of the driver's seat is where I leave my black boots when it is time for bed.

Front passenger seat

To the left of Reception is the workbench where I keep my Quendant making kit. Here, amongst the Quenns and stainless steel wire, I have a dremmel stored in a (petroleum plastic) tub held in place by the seatbelt. A dremmel is a small drill used for polishing and grinding freshly cut Quenns. This equipment is kept close to hand for any eventuality - you just never know when you will take a fancy to dremmelling a laser-cut Quenn.

My physical diary also resides on this seat, close to hand, helping to keep my thoughts and plans organised.

Tucked down the side of the seat, next to the centre console is an *Elaine Page* vinyl LP album with a picture of my eye glued to it – but that's a story for another book.

The front passenger seat is also a good place for storing snacks, like cashew nuts, bananas, and Turkish bread.

Front passenger seat foot-well.

This is where my backpack is stored. Inside the bag lives the laptop computer, with the power connector facing up, so that it can be plugged in to the juice. The bag also contains all the cables and things that you need, plus books, business cards and general stationary stuff. The bag is like a mini office. The bag contains a copy of the book titled: *The Practical Science of Planetary Medicine: GAIA. James Lovelock creator of Gaia Theory.* This is the closest thing to a holy book in the Vitan Religion.

Down here is also the folder full of music. The Gen II Prius has a CD player and none of that fancy stuff like bluetooth or whatever, so you'll want to take advantage of the CD player to play some cool tunes. Get hold of one of those CD cases that holds 50 or more discs - the ones that force you to abandon the disc box that the CD comes in. On your journeys, routinely visit opportunity shops and browse through the CD collections therein. You'll find all manner of amazing music. So fill up that bag with cool music that you get for $1 per disc.

Also in the front passenger seat well is the 'tip' where all the miscellaneous junk and rubbish gets tossed until it is distributed into landfill or recycle bins. Note: without due attention, this area can get pretty skanky, so watch out for nesting rats, cockroaches and other vermin.

If you are ever fortunate (or unfortunate) enough need to transport a guest in Motel Prius you might find that front passenger seat is just too busy, in which case, you'll need to put them in the seat behind. If you decide to seat your passenger in the front passenger seat, then you'll need to relocate all the clobber to other parts of the vehicle. Most of it can be easily moved around, but 'tip' can be a bit messy and embarrassing to relocate. This is another reason why 'tip' ought not be allowed to get out of control.

Glove box

The glove box is a royal pain in the ass – I've never been able to master it – and try not to open it too often. This is the natural resting place of paperwork such as receipts and speeding fines, dri.bots and anything else that simply doesn't have another rightful home. More discipline is required on my part, and I hope you will do better than I have.

Passenger-side Dashboard Storage Hatch

Above the glove box is another hatch that most people in the world don't know exist. Here, I store bottle of Vitan Hair Oil and a comb, so that I can look sharp at a moment's notice. Vitan Hair oil is a wondrous brew that finds many uses. It is made of equal measures of coconut oil and olive oil.

There is also a bottle of Covid-busting hand sanitiser with eucalyptus oil, that makes me swoon whenever I dope my hands with it, due to the very high dose of ethyl-alcohol and the pungent aroma of gum leaves. I try not to apply the hand sanitizer when I am driving, for safety reasons.

Centre Consul

When seated in Reception, under one's left elbow is the centre consul, inside which is a storage space with a cigarette lighter portal that allows a 12V DC outlet. Inside this space I have located a sine wave inverter that turns 12V DC into 240V AC. Into this is plugged either the laptop computer or the dremmel.

Rear left-hand passenger seat

This is general storage, where the shower bag resides and bulky, warm clothes like the waistcoat and grey jacket. Also, for some reason, if ever you lose anything, it will probably be found here. I can't explain the science behind this, but it is very consistent observation.

In the foot-well of this seat are two bags representing pre- and post-laundry. The plastic bags contain, a) clean clothes, b) dirty clothes. It is probably a good idea to have two different coloured bags to ensure that you don't accidentally take from the wrong bag. Both my bags are the same colour, but I have been lucky, so far.

Driver Side Back Passenger Seat

The back of the seat is folded forward to create the upper-body part of the bunk. This is where my torso and head resides for around eight hours a day, so it's worth making this comfortable. There are two latches on this surface, used for connecting the straps for baby seats. These devices are proud of the surface and uncomfortable to lie on, so they need to be covered with padding.

Driver Side Back Seat Footwell

Here resides a small case containing additional clothing resources, and other important items that aren't needed on a daily basis. This case also fills the gap upon which the pillows are stored.

Left hand back of hatch

This part of Motel Prius is like the tool shed. Here is a toolbox, storage for cooking utensils, and bits and pieces.

Under the floor of the hatch

Under the hatch is the butchered box with the deflated spare tyre poking through it. This is a general storage area for things that needn't be accessed regularly. This is where I stockpile all the excess bags of dri.bots: bags and bags and bags of them. Importantly, here is also stored the 12 volt air pump and the tools for replacing tyres.

There are other storage areas, such as in the door panels and under the seats, but the ones described here are the main ones.

Ballina Golf Club

On my journey, I visited the Ballina Golf Club, a vast, grassy place punctuated by big trees. It was quite beautiful, on this sunny day. I parked Motel Prius, and my first port of call was the golf shop on the ground floor. Not being much into golf, I decided to acquire an interest for a little while to see where that led. Inside the shop, I observed two listless Moderns and all manner of paraphernalia that was completely useless in preventing the collapse of the global ecosphere.

Now, golf is extremely complicated because the sport is practiced in a venue called a *golf club*, using a tool called a *golf club*. The part of the golf club (tool) that hits the ball is called the *club head*. And the manager of the golf club (venue) can also be called the *club head* – but don't refer to them as a *clubhead*, they don't like that.

If you are confused, that's okay, because it's confusing, but try to follow along, anyway.

Outside the shop I found something that piqued my interest: a barrel with a dozen second-hand golf clubs poking out the top. Intrigued, I withdrew one of the clubs and studied it with interest.

The gold club had a black rubber handle (called a grip) with the word *Slazenger* imprinted on the side. I remember the word *Slazenger* from when I used to avoid playing football, as a child.

Further down the club was a carbon fibre rod connected to a stainless steel tube, connected to a black bulbous lump – the club head. The club head had a peculiar shape, unique to golf clubs: lots of smooth, rounded surfaces, with a single flat, metallic face. On the top of the club head (the *crown*) embossed into the metal was the word, MAXFLI: a simple word that I assumed to be an aspiration for 'maximum flight'.

Most of the club head was a black, lustrous material, but the face and the *sole* (the lower face) were made of a metallic substance: perhaps titanium. The following words were written on the sole:

- *10º* (the angle from vertical of the club face).

- *Pro Series No 2* (not sure what this means – probably nothing)

- *F.H.T. Fused Hyper Titanium* (One could only guess that hyper titanium is a fancy way of saying 'contains titanium'.)

- *320cc* (the interior or exterior volume of the club head).

I don't know what came over me, but I found myself compelled to own this useless device. I went inside and interacted with the listless Modern behind the counter, and a few minutes, a few dollars, and a few hundred kilobytes of Big Data later, I exited that dour place, and put 'my' golf club into Motel Prius.

I pondered the golf club over the subsequent hours, and that night took the dremmel to it.

In the dremmel kit were these little brown discs that turned out to be quite proficient at cutting through the shiny silver metal tube that is referred to as the *ferrule*. With the dremmel, I cut the club head off the golf club that I bought in the golf club.

I smoothed off the edges of the cut metal with a different dremmel bit, and then spent some time pondering the golf club, now reduced to what could be described as a hiking cane and a hollow, titanium liquid-flask.

Motel Prius

The whole idea of 'anti-synergy' came back to mind. This is a different type of anti-synergy. When the club and the stick are fused together, they are a golf club with a singular mission: to pointlessly move golf balls from Point A to Point B, while the biosphere dies of neglect.

However, when a stick and a club are separated, the stick and head suddenly have the opportunity to become many useful things. For example, the stick becomes a hiking cane, a thrashing or poking weapon, and a pointer. And the club becomes an incense holder, a penholder, a whisky or Vitan Oil flask (320cc). Maybe, if filled with bee's wax and a wick, it could also become a candle. *Hmmm.* The possibilities are endless.

I sat in Reception in Motel Prius and pondered the severed head of the former golf club. I noticed that it was a bit grubby, so I retrieved a dri.bot, but when I pulled one, three came out, all tangled together. I cursed at the terrible waste of plastic, and then washed the club until it was shiny.

As I sat in Reception I listened to one of the CDs that I had bought for $1 from an opportunity shop - *Antarctica by Vangelis* - and I pondered a hypothetical conversation at a market stall where a Vitan Minister offers the club for sale for $300 to help fund the cause.

A male Modern walks up to the stall and opens his mouth: *"Hey, that's a golf club."*

The Vitan Minister shakes his head, wearily and replies: *"This is a flask in the shape of a golf club that can be used for storing liquids, such as whiskey or Vitan Hair Oil."*

Modern Man: *"But it looks like a golf club."*

Motel Prius

Vitan Minister: *"I have already conceded that."*

Modern Man: *"Serious, mate. That's a head off an old golf club."*

Vitan Minister: *"I know its lineage. I decapitated it with a dremmel, last night."*

Modern Man: *"You can get those for $5 at the Ballina Golf Club."*

Vitan Minister: *"Ahh! But can you get a golf club-shaped whiskey flask at the Ballina Golf Club?"*

Modern Man: *"Nahh. They don't sell that sort of stuff."*

Vitan Minister: *"Exactly. So, cutting the club off a golf club - in a golf club - it's sacrilegious, isn't it?"*

Modern Man: "What does that mean?"

Vitan Minister: *"It means that it's wrong. So this is the only place where you can buy this. Are you paying cash or card?"*

Modern Man: "But it's like $300."

Vitan Minister: "Sure. I get that. But let's frame it this way. If you buy this you'll be helping to advance the Vitan Religion. And if you don't buy it, there will be some ecologically ruinous opportunity cost, with you going out and buying some petroleum-based shite that will shortly end up in a landfill."

Modern Man: "But what will I tell my girlfriend? She checks all my transactions."

Vitan Minister: "Tell her that it is an egregious crime that titanium should be put to such poor use, only to be cast aside into the refuse stream, shortly thereafter. Tell her that you bought this to support **Imperium Vitae-planeta**, and to Advance the Verdant Age, the time when human civilization and the Living Planet thrive in synergy, deep into the Long Future."

Motel Prius

Modern Man: *"I don't even know what that means."*

Vitan Minister: *"That's okay. Because once you have tried to repeat it to your girlfriend, there's a good chance that she won't either. It's not important that you understand, just that you move and inspire her. I'll tell you what I'll do. You give me three hundred dollars folding money, and I'll give you the liquid holder that looks like a golf club, and I'll throw in the hiking stick as a bonus."*

The Vitan Minister leans towards the Modern Man, and whispers: *"And because we are having this intimate conversation about Vitae-planeta, for an extra hundred dollars I'll throw in a Quendant."* He pops the Quenn necklace from around his neck, and presses it into the hands of the Modern. *"Give this to your girlfriend. Tell her that it is a powerful symbol that helps put relationships where they ought to be."*

Modern Man: (Staring at the Quenn in his hand). *"This is like some sort of witch-doctor shit, isn't it?"*

Vitan Minister: *"Absolutely. The automatic teller machine is over there. Four hundred dollars."*

Now, I am not sure that a conversation like that would ever occur, but it could.

As a Vitan Minister it is important to master the concept of *Inventing Possibility*. Create in your mind an unlikely and yet positive situation, and then take action to bring it into being by enrolling people into the idea.

Here's an example: the biosphere is dying, and all the indicators suggest that it will collapse in an overheated, acidic mass sometime in the coming decades due to ocean acidification and a cascade of climate tipping points. This is the logical conclusion of humans continually growing the global economy based on fossil fuels and the destruction of nature.

If, on the other hand, we collectively invent the possibility of stabilizing the global economy based on renewable energy and the regeneration of nature, we would likely end up in a happier place: maybe even the Verdant Age.

It's important to realise that you don't need all the people to be aboard such a scenario for it to come to pass.

You only need a small proportion of society to shift the trajectory of that society. And you only need a small proportion of societies to create the path for the rest of the world to follow.

One reason why the transition to clean energy hasn't happened yet is that no one has gone first. Sure, Costa Rica might have gone first, but Costa Rica doesn't play with the big boys. We need the UK, or Germany, or France, or Japan to commit to rapid decarbonisation to create the space for other enlightened nations to follow. Once this is underway, then China, Russia, USA will come on the journey, too.

Just one brave nation, is all we need to begin. That's when we start to see a global movement to avert human extinction.

It hasn't happened yet in part because the so-called 'free countries' aren't free from the wealthy people hijacking the democratic systems to suit their own interests.

"Oh, but money always infects politics," I hear someone say. *"The political class always play second fiddle to the merchant class…"*

Sure. Maybe. Whatever.

That's not the point. In the model called *Inventing Possibility* the past and future are irrelevant. What's relevant is *the now and the possibility*. And the way that you implement the possibility is by enrolling people in the possibility.

The people who founded the *Extinction Rebellion* knew this. They researched non-violent rebellions over the past 200 years and found that none had failed when 3.5% of the population supported the rebellion. And while the non-violent rebellions played out on the streets, the political class fought each other, the police and mainstream media ran opposition, and Mr and Ms Jo Public sat in their lounge rooms watching the rebellion play out on the TV, not having a firm opinion about whether the rebellion was a good idea or not.

The problem with climate change and ecological collapse as an issue to underpin a rebellion is that neither has become apparent to the average Joe. Average Joe just isn't that bright or interested enough to realise that something is going on. So getting 3.5%

of the public enthused to organise a non-violent rebellion is hard, right now.

The dilemma is that by the time food becomes scarce in the Western countries due to multi-bread basket failure – and Average Joe wakes up – it maybe too late to do anything meaningful, as societies will collapse due to food shortages.

Multi-bread basket failure is a fancy term that describes the key food production areas being unable to deliver their harvest because of extreme weather events (e.g. heat waves, flash drought, flood, hurricane) or political strife (e.g. war prevents agricultural activity).

Multi-breadbasket failure is writ large in the coming decades. One science paper says that 600 million people will be exposed to fatal temperatures by 2070 in the central China food bowl due to Wet Bulb 35 temperatures. Wet

Bulb 35 is a ratio of temperature & humidity, beyond which humans fall dead in hours due to heat stroke.

My friend from the Sunshine Coast described how a heatwave affected her property. It was over 40 Celsius for days, and on this particular day, about 45 C. She describes how she heard this ghastly sound at the back of the shed and walked around, sweating profusely, to see one of the bushes withering in the heat. The plant was exuding some sort of gas or vapour and making a whining, hissing, howling noise: *as though the plant was screaming in pain.*

The word *Anthropocentric* comes to mind. It's another one of those *Anthro-* words, which suggests that it is about humans. Anthropocentric means placing humans in the centre of things, in the centre of our concern. Here we are writing papers about how climate change is going to affect the wellbeing of humans. But, where's the paper about how the climate change (that we caused) is hurting the plankton, or fish, or river dolphins, or oak trees, or the little mice that live in the hedge rows…

We humans have spent the last two hundred years burning fossil fuels, butchering indigenous cultures, carving up nature, and destroying the biosphere, and all we can talk about is how this is going to hurt us.

Unsustainable Super-predators only think of themselves.

It is perhaps not surprising that many environmentalists are misanthropes. What's that? To quote the movie *The Proposition,* a misanthrope is, "*Some bugger that hates every other fucking bugger.*" A misanthrope is someone who hates people. Many people see humans as a virus on Planet Earth. It's easy to see why they think that.

Motel Prius

There's an excellent line in the movie *Matrix* where *Agent Smith* (Hugo Weaving) lectures *Morpheus* (Laurence Fishburne) who is bleeding and tied to a chair. Agent Smith says, *"I'd like to share a revelation that I've had during my time here. It came to me when I tried to classify your species. I realised that you are not actually mammals.*

Every mammal on this planet instinctively develops a natural equilibrium with the surrounding environment, but you humans do not.

You move to an area and you multiply, and multiply, until every natural resource is consumed. The only way you can survive is to spread to another area.

There is another organism on this planet that follows the same pattern. Do you know what it is?

A virus. Human beings are a disease, a cancer of this planet; you are a plague.

And we are the cure."

This is not an uncommon view amongst Moderns and Cultural Creatives, alike. However, this is not Vita's view.

Vita doesn't view humans as a plague, even though we act like one. *Vita sees humans as an organism that simply isn't doing its job.*

All natural life forms on Earth have important roles in helping to maintain the function of *Vitae-planeta*. And *Vitae-planeta's* function is to maintain biosphere homeostasis. Homeostasis means to maintain conditions within the range that allows life to prosper. We support her, and she supports us. It's simples.

As an example of homeostasis, in the human body are mechanisms that keep core temperatures within a narrow range. The same goes for the pH of the blood, and other key parameters.

Motel Prius

This is the role of all species, to find a balance in the ecosystem by which they can assist to maintain conditions on Earth suited to the wellbeing of life on Earth.

That responsibility extends to us humans. The problem is that we humans – as a result of our free will, ignorance and cultural practices - neglect our duties.

Let's go back to *The Matrix* movie for a while. In the story, human beings have been duped into believing they inhabit the real world. However, in reality, they are actually curled up, fast asleep, in vats of goo, and their bodies are being probed for nutrients and energy by machine robots of the future. The world that the sleepers perceive is actually a complex holographic dream installed in their minds by the machine overlords, to keep the sleepers calm.

In the story, *Neo* (Keanu Reeves) lives in the hologram dream. *Neo* is approached by *Morpheus*, who has hacked into the Matrix to coax *Neo* out of his slumber. Morpheus offers him the choice of taking either a Red or a Blue Pill. The Red Pill is the pathway to waking from the hologram, but entering the real world of perpetual war against the machines.

Neo takes the Red Pill and wakes in the vat of goo. If you want to see the visuals, search for images or videos using the term *"matrix neo wakes"*.

Vita is a new religious movement that seeks to help people wake to the *climate and ecological matrix*.

We want you to wake up to the warm, acidic goo. We want you to *'take a look'* and *'be careful'*.

To be clear, Vita does not seek to wake you to your *inner-self*, your *seventh chakra*, or your *true earning potential*. These are all good things, maybe, but that's not Vita's game.

Vita helps to connect you to established mainstream Earth Science and the political science that can help us get out of this shitty, fucked up mess before it's too late. Vita does this by grounding your spirituality to *Vitaeplaneta*, the Living Planet, and helping you to develop the capacity for *Big Talk*.

Vita seeks to enrol 53 million people across the Western World to invent the possibility that the biosphere doesn't die, and that the humans and much of what remains of nature gets to live deep into the Long Future.

It's a big task. Enrolling 53 million people - a quarter of the Cultural Creatives in the Western World – and having these people identify as Vitans, and be open to what Vita has to say.

I think it is eminently doable. But maybe that's because of my disposition. Maybe I am not paying attention to details. Maybe I am just a dreamer.

If you think that it seems an impossible task, then I suggest you go buy yourself a dremmel and some second hand golf clubs and rent yourself a market stall. Why? Because if you can't see the possibility of something big, maybe you need to start working on the possibility of something small.

Work your way up from there.

Whatever you do, just take the red pill and wake up in the goo. Then take a look, but be careful.

Tear Down The Statues

The city of Townsville, where my chest resided before the journey, had been my home for 11 years. In the middle of town, there was a bronze statue of a guy called Mr Towns who was a founding figure in the establishment of the city. I glanced at that statue a few times, but never really thought much of it.

So it turns out, Mr Towns was involved in the mass importation of poorly paid and poorly looked-after labour for his various enterprises. He palled-around with blackbirders and slavers, and was probably a slaver himself. These sorts of statues are everywhere, commemorating the people who founded towns by subjugating people – people, whose ancestors are alive today, and are rightly affronted by the insensitivity of the statues. If the Robert Towns statue had the following written: "*Robert*

Towns was an abusive sociopath who enslaved thousands of people of colour for his own gain, but this was morally acceptable to the white people of the day," at least it would go some way to speaking the truth.

And then a Caucasian-American cop knelt the Afro-American citizen to death, and a wave of rage against institutionalised racism swept the world. One way that this expressed itself was in the toppling of statues.

At time of writing, the statue of Robert Towns remained intact although some cheeky rebels has spray painted its hands painted red. Mr Towns got off lightly.

I wish I had been more enlightened, but in hindsight, I wasn't. I have only recently understood that we are surrounded by statues that adulate some really foul human beings.

Motel Prius

Here's an example. I was taught at school that a guy called Christopher Columbus sailed a boat and came across some islands, and that this discovery was a significant part of my cultural heritage. I wasn't taught that Christopher Columbus was an enabler of mass child rape, as he proudly wrote in his own diaries.

What's that all about? Were my teachers ignorant of the foulness of this man? Or were they in on some big act, to make okay the wickedness?

A good way to make sense of all this madness is to consider the idea of *metanarrative*. Narrative is a story. Metanarrative is the grand story that creates the context to all the sub-stories that follow. Here is an example of a metanarrative: risk-taking entrepreneur gets a boat, discovers new lands, finds treasure, and it's all good for the motherland. Let's make a statue of him.

If this metanarrative is accepted by the masses, then all the sub-narratives are accepted without question. Well, guess what. The metanarrative is being dismantled one racist statue at a time.

Vita's metanarrative is much more sensible and life friendly than that which we are accustomed today. Simply put, Vita sees humans as just another organism, whose most outstanding feature, is that we are not doing our job of nurturing Vitae-planeta. If we don't drive ourselves extinct, we may be able do learn to do our jobs, and help Vitae-planeta prosper. In so doing, we enter the Verdant Age, where humans and nature can coexist for hundreds of millions of years.

In order for this new metanarrative to take effect, it is necessary that we shine a light on many of the muddle-headed ideas that society accommodates without question.

Yoga

When I arrived in Byron Bay, the end of my journey, I found myself in a world of yoga. I'm reminded of a joke I saw on facebook a while ago. It was a single image cartoon with two speech bubbles. At a funeral service, the master of ceremonies asks the gathered crowd, *"Would anyone like to say a few words?"* A solitary hand raises, and the person says, *"I'm Vegan."*

I am still laughing at that joke, as I write my own version of if.

A Vitan Minister addresses an audience and says, *"Does anyone have insights into ecological sustainability?"* A solitary hand raises, and the person says, *"I do yoga."*

Now, I don't have anything against yoga – in fact, I would like to learn yoga – but I am concerned that much of what is said about *yoga & sustainability* is just bunkum.

You'll hear people say things like, *"Oh, sustainability, I'm into that. I practice yoga."*

And I'm like, *"Ahh? So, how does that work, exactly?"*

Yoga Person: *"Yoga grounds you, connects you to the Earth."*

Vitan Minister: *"Really? But you are sitting on a polyurethane mat and wearing spandex pants; and the mat and your designer-wear are made of petroleum oil, which is kind of like the Earth problem. And you walked here in shoes with synthetic rubber soles, shedding micro-plastic particles into the soil, or you drove here in a petrol-powered car. So, how is that grounding you to the Earth? I don't get it."*

Yoga Person: *"I think that we are talking about different forms of sustainability."*

Motel Prius

Vitan Minister: *"Okay, so which one are you talking about? Because the one I am talking about relies on maintaining the habitability of Earth's biosphere."*

Yoga Person: *"I'm talking about soul work, about being in balance."*

Vitan Minister: *"In balance with what? Gravity? You learned that when you were two."*

Yoga Person: *"In balance with nature."*

Vitan Minister: *"I don't think that you are hearing me. Sitting on a piece of plastic made in China, wearing plastic pantaloons made in China doesn't balance anything ecologically; no matter what angle you hold your legs. I'm talking about the intellectual discipline in understanding and accepting the* **'priority of ecological rationality'***, which, simply, means that you don't get to practice yoga on a dead planet. And a dead planet, as far as we humans are concerned, is a planet where the global average temperatures are half a* *degree Celsius or more above pre-industrial baseline temperatures. More than that and we risk heading into the Hothouse, and taking down most of nature, like a rerun of the Permian Extinction. Yoga maybe good for your mental health, but it is immaterial for the habitability of the Home Planet."*

It is about this point of the conversation where the yoga person either storms off in a huff, or starts to eye-off your Quendant. Experienced Vitan Ministers will know what to do in these circumstances.

I want to take a step back here. I know that I seem harsh, when I am simply trying to be insightful.

Much of what is discussed in the frame of spirituality is meaningless with respect to keeping humans alive on this planet.

This is important, so I will repeat it in italics: *Much of what is discussed in the frame of spirituality is meaningless with respect to keeping humans alive on this planet.*

There is all this talk about Heaven & Earth, Cosmic and Universal Energy, crystals and dream-catchers, and Near Term Human Extinction.

There are lots of ideas that are distractions to a nature-based spirituality that can lead to the *Long-term Extantion* of the Human Race.

Take a moment to ponder that odd word: *Extantion*. It is the opposite to extinction. So, *Long-term Extantion* is the opposite to the often-described *Near-term Extinction* of the human race. There are a lot of people convinced that the extinction of the humans is inevitable. Vitans don't hold that view.

Another concept that needs attention is *Heaven & Earth*. We need to let this idea go, right now. It is so totally past it's use-by date.

One of Vita's canons of conduct is to *Reinvent New Year*. This is an idea where we rethink the institutions and ideas that we were born into. We need to let go of the word heaven. It is not a word that is conducive to fostering right action for sustainability.

We humans are biological components of Earth's biosphere. We grow here, die here, and our remains enter the biophysical flux on this planet. Be content with that. It's enough.

Plus, we needn't concern ourselves with cosmic or universal energy. Instead, focus on solar energy that powers photosynthesis.

Earth has a magnetic field that protects life on Earth from cosmic energy – cosmic radiation and sub-atomic particles ejected from the Sun. We needn't worry about what happens energetically in other parts of the universe.

Motel Prius

For Vitans, our circles of concern are just two fold: the circle of the Earth rotating around the Sun, and the circle of the Moon rotating around the Earth. Everything else in the solar system, galaxy, cosmos and universe is completely immaterial to our wellbeing.

Don't get me wrong, if yoga is a good thing for you, keep doing it. But being happy because of yoga is different to being a positive participant of the habitability of Earth's biosphere.

It is possible, but not desirable, that humans could live on this planet completely miserable, but ecologically sustainable. Consider the European Dark Ages: disease, cold, wealth inequality, the excesses of the Church, the Inquisition. And yet in that era, the humans were not such a force of nature as to perturb the ecological balance. The Dark Ages were probably ecologically sustainable.

Consider another age where humans are happy and peppy and yet pumping out heat trapping gases that triggers the Hothouse. Imagine wearing spandex and driving a 4WD vehicle to a yoga session. Being happy and having good mental health is not the same thing as ecological sustainability.

Vita seeks that we can have both. But this requires an intellectual discipline. So, let's keep it simple and focus on the important things, and not get distracted by things that don't make a difference.

Let's develop a disciplined spirituality that leads to the right action that advances the longevity of humans on this planet.

If we get this right, we may yet be practising yoga 50 or 100 million years from now. As an aside, I dare say that in 50 million years, there won't be any spandex yoga pants.

Motel Prius

The Boulder

There is an imaginary town with a beautiful beach beneath a long cliff. At one end of the beach is the town, and at the other end a popular café district. While there is a road and footpath along the top of the cliff, the accepted custom to get from the town to the cafés is to walk along the beach. This is a daily ritual for most of the townsfolk.

One day, a huge storm dislodges a massive boulder from the top of the cliff that falls onto the beach about halfway between the cafés and the town. The rock lands so close to the cliff there is a gap of less than a meter between its side and the cliff.

On the morning after the storm, the townsfolk gather at the boulder and ponder this impediment to their normal route. Someone suggests moving the boulder, and an engineer arrives with a measuring tape and calculates that the boulder isn't going anywhere. It is simply too heavy to drag, and blasting it into pieces would likely make the cliff collapse.

Eventually, the townsfolk adapt to this new circumstance, and find many ways of traversing the boulder.

- Hug it and inch your way around it, clung to its surface like a crab.

- Run at it and leap, and in three bounds you will be atop it and then on the other side.

- Squeeze your self between the boulder and the cliff.

- Wait until low tide then walk around the seaward side

- Wait until a very high tide then swim or row a boat over it.

Despite all the options, no one has yet found a way for the boulder *not to exist*. The boulder is an objective reality. It's tangible. Physical.

Some people imbibe drugs or incite spirits or mystical forces, and in this tranced state, they perceive the boulder to no longer be there. However, in the cold light of day, after the session has worn off, the boulder remains.

Some people simply don't talk about the boulder, won't even permit it to be discussed in their homes, as though avoiding mentioning it makes it go away. But even those people, when walking along the beach alone, still find themselves confounded and having to choose one of the established options for moving beyond the boulder.

In the early days, some people used to go up to the boulder and scream profanities at it, basically telling the boulder to *fuck off back to where it came from*! But the boulder gave no indication of having even heard them.

On and on, I could go about the boulder, but the point of the story is that there is a thing in the world call *objective reality*.

You may not like to hear people talking about ecosystem collapse; you may not believe that climate change is real or a problem; you might think you have found a way to tunnel under ocean acidification... The point of the matter is that the scientific method deployed throughout the world's leading scientific institutions is telling us that we are heading towards a global calamity that will snuff-out human civilization, and likely kill off most species on Earth. Just google the phrase: *scientist's warning*' and read what they have been saying since the mid 1970s.

If you want to know what to wear on Friday night, use the decision make tool called 'whim'. If you want to know whether you should drink ten beers and a bottle of bourbon, check with the decision making tool called Doctor's

Advice. If you want to know what the global average temperature of Planet Earth will be if we continue to dump 30 Gigatons of CO_2 into the atmosphere every year, don't ask Whim or the Doctor. Ask an Earth System Scientist. It's a simple principle. Science is shithouse at advising what you should wear on Friday night, or whether you should have chicken or fish on an airliner. But when it comes to understanding the physical world – chemistry, physics, oceanography etc - it is the only source of truth we can count on.

At some point, society needs to acknowledge the boulder on the beach and act appropriately. And what that means is to bring about the political conditions that foster drawing down three trillion tons of CO_2 from the atmosphere and winding-up the entire global fossil-fuel industry, by mid-century. That is a very big boulder, indeed, and the sooner that we wake up to the task, and get to work, the more likely the ecosphere won't collapse around, us as we try to set things right.

The Northern Rivers

When I was in Townsville, I wrapped up nine boxes of personal possessions and sent them by road freight to my friends in the Northern Rivers. The freight company messed up, which was no surprise to me, and the goods ended up stranded in a Lismore depot. I drove down the coast from Townsville with the Sea Chest in the back of Motel Prius. For logistical reasons, I dropped the Sea Chest off in Brisbane, then drove to Lismore and moved the nine boxes to a storage area with my friends.

With the gear safely stored, I then set about trying to find a place to store myself. I was new to this area, and had much to discover, as well as setting about trying to create some micro-enterprises to make a living. I scouted around for somewhere to park Motel Prius, and found the best deal was the

Motel Prius

Broken Head Caravan Park, about twenty minutes south of Byron Bay.

I have been living in Motel Prius in this place for three weeks, at time of writing, and I love it. I wake to the sound of rain or bush turkeys jumping on the roof. I open my eyes and look out to the Pacific Ocean. I can hear the surf from the car when I turn the music down.

My eyes see beauty, and yet tucked away in a recess in my heart is a profound sadness. That beautiful ocean is acidifying as it soaks up the carbon dioxide gas emitted by the human economy. So few people in this paradise have any concept of Earth System Science, and what is coming down the pipeline under climate and ecological collapse. This last summer in Australia, a quarter of the nation's forest burned in intense wildfires, and half of the formerly-Great Barrier Reef bleached from excess heat in the Coral

Sea. It was recently 38C in the Arctic! The whole system is coming down, unless we humans intervene to reverse it.

I know what I sound like to most people: the Chicken Little character saying, *"The sky is falling"*. But, it's not falling, it's burning.

Just today, I had a coffee at the Treehouse at Belongil Beach, and then wandered down onto the sand. This is a dog beach and I watched these animals go crazy, running at full speed along the wet sand and play fighting in the surf. Maybe it was the vestigial traces of the psylocibin that I may or may not have consumed, that made the light playing in the ripples of sand and the small wavelets washing over them seem so scintillating. This really is an extraordinary place. And there are so many extraordinary places where one can feel a deep connectivity to the scenery, the feel of sunlight on your face, the white noise of moving water, the sense that all is well in the world.

But for me, there is something sinister about all this beauty and wonder. It is deceptive. It's a distraction.

Regular folk can't see through this wondrous view. But I can. I read the Trajectories science paper (*Trajectories of the Earth System in the Anthropocene*) and I talked to the lead author, and recorded the interview on my Perium youtube channel.

I trust Earth System Science. And what these scientists say scares the shit out of me.

The scientists say that the global average temperature gets somewhere between 1 & 2 degrees Celsius about the pre-industrial baseline temperatures, the Earth System triggers *a cascade of climate tipping points* that drives global temperatures into the Hothouse. Then, we all die, and the mammals all die, and the ocean dies, and all that's left is the funny looking fish at the bottom of the sea.

Somewhere between 1 & 2 degrees, all this tragedy pours onto our heads.

And we are already at 1.3.

The Red Cats

In that fictional town with the boulder, is a fictional woman called Debra who had a fixation with Red Cats.

I said to her, one day, *"What's with the Red Cats? You've got, like five of them."*

"They make great pets," she said.

Vitan Minister: *"But, there's many different types of pets. There's dozens of breeds of birds, cats, dogs. There's fancy rats, insects, lizards. Some people even keep pangolins! Why have you fixated over one particular breed of cat, five times over?"*

Debra: *"Because they are beautiful."*

Vitan Minister: *"But so are Praying Mantis and pugs."*

Debra: *"And they talk to you, and jump from place to place."*

Vitan Minister: *"But so do parrots and crickets. Tell me something. When did this cat thing begin with you?"*

Debra: *"When I was five, I spent a week with my aunt, and she had Red Cat. It talked to me, and jumped from place to place around me. I never forgot it. And when I moved into this house, I knew I had to have one."*

Vitan Minister: (Looking at all the Red Cats jumping from place to place.) *"One?"*

Debra: *"Well, it started with one."*

The single interaction with a Red Cat had so impressed young Debra that from that point on, she projected sacred values onto Red Cats.

She couldn't explain rationally why she loved these cats so much, except that she had met them when she was young. They had inched their way into her psyche to such an extent that she could not conceive that there was anything peculiar about her affection for them.

To be clear, there was nothing peculiar about her affection for the Red Cats. People fixate on things all the time, in a manner that cannot be understood through reason. It's a *spiritual thing*, and humans do it a lot.

Because Debra ascribed sacred values to Red Cats, she adopted behaviours (some might call them *canons of conduct*) that bought her belief into life. She owned Red Cats, she nurtured them, she played with them, groomed them, made sure they had proper health. And when one of the Red Cats escaped the house, she fretted to the point of exasperation.

In all these regards, Debra has a religious fixation with the Red Cats.

Had Debra done the paperwork, and submitted to the Charity Commission, she may even have been able to create a *registered religious institution* devoted to Red Cats. In this way, she could have created a structured pathway for other people to honour and nurture Red Cats, and in turn be nurtured by them.

"Oh, don't get caught up with tabbies or those black Persians," her Red Cat Religion website might say, *"They don't talk or jump from place to place."*

So it is with Vita. Vita is a structured pathway for people to develop a spiritual connection to the Living Planet, and for them to be nurtured by the Living Planet.

For this journey to begin, it is necessary to *meet the Red Cat* – to have an emotional or intellectual experience that forms the spiritual bond to the *Living Planet*. Vitans call this Ecophany: *ecological epiphany*.

I don't know what *meeting the Red Cat* entails for you. Maybe it is seeing hundreds of sulphur-crested cockatoos alight from a tree at sunset, or a Full Moon rising across a lake. Maybe it is seeing the full depth of the cosmos on a clear night in a place untouched by artificial night sky brightness. Maybe you need to have a near death experience to awaken to the ongoing near death experience that the *Living Planet* is suffering at the hands of human civilisation on a daily basis.

Maybe you need to watch a movie about the annihilation of Hiroshima with a small nuclear bomb, and then wake up to the fact that there are 17,000 big nuclear bombs today, and that ignorant sociopaths like Trump have the keys.

Personally, my *Red Cat meeting* came when I saw the plastic trash in the ocean off the coast of Taiwan from the bow of an oil exploration ship, when I was aged 23. I know a woman who's Red Cat moment came when she flew over Greenland, and saw that it was melting. All environmentally minded people cant recount the moment that they awoke and started taking action.

If you still haven't accepted *Vitae-planeta* as the locus of your spiritual belief, I invite you to find a way to do so. *Find your Red Cat.*

We are out of time. There are too many Moderns chomping through the world's natural resources. We need more Vitans. And some Vitans will go on to become Vitan Ministers, and some of them will adapt a Gen II Prius into Motel Prius, and go on a spiritual journey to share Vita.

Maybe that is you.

Motel Prius

The Happy Ending

This story has covered a lot of ground since the flat cane toads in the opening chapter. As a story, it is quite clearly lacking some of the normal structures that picky book reads generally expect. While it has a protagonist of sorts – the Vitan Minister – it lacks an antagonist, character arcs, three-act structure, and all the normal goodies that one might ask of a novel. But that's okay, because it's not a novel. It's not even a stream of consciousness. It's a dump of ideas, or a *Declaration of Independence*, or a cry for help, maybe.

In truth, I don't actually know what this book is, *and I wrote it*.

Let's just call it a *Covid Story*. Everyone will know exactly what that means.

But even a *Covid Story* deserves a happy ending. Let's call that euthanasia, for want of a better word. This chapter is a self-imposed mercy killing, to give this story a peaceful end.

Our planet is dying. Act accordingly.

You can't honour that statement on a yoga mat or a surfboard. At some point, you have to get your head around Earth System Science, and take appropriate action. The practice of a *sustainable religion* – a religion devoted to ecological sustainability - can be an important part of this action.

People freak about when they hear the word religion. *"I don't like religion,"* they say, even though they happily practice multiple religions, every day.

My hypothetical friend believed that Red Cats were sacred, and she acted accordingly. The fundamental difference between her cat spirituality and a religion, like Christianity, is the existence of an institution.

There are *institutional religions* (Christianity, Hinduism, Islam, etc.)

and non-*institutional religions* (Red Cats, Surfing, Yoga, Football, Holden Commodores, etc.).

One type of non-institutional religion holds that nature is sacred, and ought to be protected.

In Australia, that religion shows up in the 5-yearly census under the headline, *Religious Affiliation: None.*

This is one of the reasons why the planet is dying. There is no official voice for those who hold a nature-based spirituality.

A full 30% of Australian's declare their religious affiliation to be 'None', even though we are a nation renown for its love of nature: the ocean, rivers, beaches, national parks… all that stuff.

So, Vita has created an institution that *acknowledges nature-based spirituality* and seeks to get the concept the respect and recognition it deserves.

For example, Vita has applied to the Bureau of Statistics to add *'nature spirituality'* to the census form. I think that about a third of the *'Nones'* would tick that box if it were available, and *just like that* a full ten per cent of the Australian public will show up in the official statistics as having a spirituality devote to the Living Planet.

This is important stuff, and it needs to happen across the entire Western world. Most Modern people have spiritual connections to mono-god religions, consumerism, and pets; and none of these things are helpful in averting ecosystem collapse. And spirituality is a powerful driver of action.

This does not mean that one who holds a nature-based spirituality needs to pay homage to Vita. It simply means that there is now a structured pathway to help regular, mainstream people to create a spiritual connection to nature.

Motel Prius

Our planet is dying, and we need to do everything in our power to prevent that. Vita Religion is a new idea that needs to be properly tested.

I wish I could claim to be some super-brainy, sagacious, fortune-telling witch-doctor type who could heal all your emotional and psychic ills just by telling you some stories. I don't claim to be a planetary healer. But, with that said, I have set about writing and publishing this book with intent. I believe that there are valuable truths in this story.

I am not saying that this book will automatically right the ecological wrongs of the world. Indeed, there is a measurable ecological burden imposed upon *Vitae-planeta* just by making this book available to the reader.

To produce the paperback, trees had to die, and there is a small amount of fossil fuel-based material associated with the cover, or the wrapping of the book, or the energy used to transport it to you.

In the electronic copy of the book, no trees are harmed, but the transmission of the ebook requires the activity of hundreds of computerised devices called 'servers' operating across the planet – all powered by electricity, most of which comes from burning fossil fuels.

My worry is that the ecological benefit that comes from telling my story is overshadowed by the ecological harm wrought by having told it.

Sure, I could travel from town to town, telling my story person to person, but it would be slow going without the massive ecological footprint of Motel Prius.

So, I am a bit stuck, and my way out of this conundrum is to invite you, the reader, to make good with this message, and see to it that all the ecological havoc wreaked by Motel Prius and the publication of this book comes to good end.

I invite you to use this book as a fulcrum, a lever to pry you from your inertia and propel you into action to *Advance the Verdant Age*.

Take a look. Be careful.

You'll need to engage the rational part of your mind, and the part of your being that senses the spirituality inherent in the natural world: *Vitae-planeta*. The engagement of your rational mind and your spiritual senses acting together forms a Vitan Synergy: a *Vynergy*.

A *Vynergy* is what results from a mind rationally engaged in Earth Science + spiritual attachment to *Vitae-planeta*. A *Vynergy* creates something quite extraordinary: an agent of change for the Verdant Age.

Covid-19 is a disease caused by the SARS 2 Coronavirus. Despite how it may sometimes seem, humans are neither a virus, nor a disease.

In the song *Clocks* on the album *Rush of Blood to the Head* by *Coldplay* ($1 at your local opportunity shop) are the words:

Come out upon my seas
Cursed missed opportunities
Am I a part of the cure?
Or am I part of the disease?

Vita can answer that question. Covid-19 is a disease. You are not. Instead, you are one of two types of living beings:

- a human being who is doing their job, nurturing *Vitae-planeta*

or

- a human being who is not.

It really is that simple.

oOo

Backmatter

About Guy Lane

Guy Lane is an Australian/UK dual-national living in Brisbane. He has a Bachelor of Science with Honours in Environmental Science and has nearly completed a Master of Business. Since a young age, Guy has been on a quest to answer the questions of *Life & Earth*. His professional life is eclectic, to say the least: from offshore oil exploration, environmental consulting to a variety of entrepreneurial ventures with sustainability themes. He is the author of multiple fiction and non-fiction works and the founder of Vita Religion.

www.guylane.com

Vita Religion

Vita is a new religious organization that was registered through the Australian Government's Charity Commission on 20 March 2020.

The word 'religion' rightly triggers a lot of emotions, but to be clear, Vita Religion is not devoted to a God, but to life on Earth. 'Vita' is the Latin word for 'Life', after-all.

Vita Religion teaches that all the living things on Earth are parts of a single life form called *Vitae-planeta* – the Living Planet. This means that we humans – you and I – are cells in the body of the Living Planet.

www.thinkvita.org

The Quendant

The Quenn is the symbol of the Anthropocene Epoch, and a Quendant is a Quenn Pendant.

The inner circles represent (a) the physical, non-living part of our planet (b) human society (c) the living systems of our world (e.g. Planetary Boundaries). The outer circle represents Earth's biosphere, the living skin on the surface of the planet. The missing section at the bottom indicates that the biosphere is less that whole, today.

The Quendant is laser cut from 2mm stainless steel, comes with a stainless steel wire and magnetic clasp.

www.thinkvita.org/the-quenn

Fiction by Guy Lane

INTERVENE – Spaceman Zem is trained to transform the global economy to make it sustainable, but when a heavy drinking Earth woman takes a shine to him, he gets right out of his depth.

THE MOOGH – The best thing that ever happened to young journalist Maggie Tarp is the Moogh, the mysterious monster who come to show humans how to live properly on Earth.

YONGALA – On the final voyage of the S.S. Yongala, an assassin befriends the niece of the man he is sent to kill in this emotionally charged tale of adventure, innocent friendship and redemption.

THE OIL PRICE – A satirical thriller set in Dubai where a wealthy playboy finds himself in the crossfire between an oil firm and the environmentalists who would stop them.

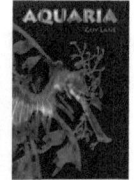

THE MARTIAN CAT – A grim, adults-only satire about a poor, diseased cat on Mars and the naïve space engineer – Charlie Darling – who would bring it back to Earth.

LOVE IN THE AGE OF BITCOIN – Action man from Bitcrime Division Absinthe Rhinohorn finds that the Chong Triad has ambushed his sweetheart Undercover Agent Turtledove.

HEART OF BONE – A taught psychological drama about a billionaire who torments his staff by manipulating their most sensitive emotions.

AQUARIA – Aquatic adventure woman Lucy Callahan dives into action when an oil company threatens the Aquaria marine science facility that she built.

www.guylane.com

Who's Next? Climate, Collapse and You.

Freelance environmental scientist Guy Lane shares insights from his 30-year quest to understand what comes next in the climate & ecological crisis.

In just two hours, this fast-paced, science-based book lays bare the grim truth about collapse, what it means for you, and how you can help make it better.

Read It. Share It. Take Action. Start Now.

www.guylane.com/whosnext

Time to Wake

Australian Reality TV star Suzi Taylor features on the cover of the Sequel to *Who's Next*.

The *Time to Wake* series seeks to provide updates on the climate and biodiversity crisis by Guy Lane in short non-fiction books released regularly.

Time to Wake: Climate Change is Here shows that now even the rich, powerful and famous are vulnerable to extreme weather events fuelled by climate change.

See Suzi Taylor's video introduction.

www.guylane.com/timetowake

Mister Bluesky

Professor Timothy Bluesky shows that sometimes, the best way to get back is to walk away.

Mister Bluesky was conceived the *Celestine Prophecy of Sustainability* - a world-changing book that 'everyone' knows of or has read.

Mister Bluesky is a manual for a spiritual and ecological revolution that introduces the *Manifesto for Life & Earth*.

www.guylane.com/misterbluesky

Death Star

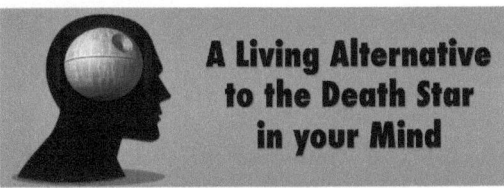

This essay introduces Vita's cultural mythology and was written prior to Vita being named.

Human civilization is blindly driving the living systems of Earth to extinction. We need new guiding principle to set things right.

www.perium.org/data/death-star.pdf

The Rebel Offset

Not everyone has the pluck to participate in civil disobedience for the climate and biodiversity crisis. Every rebel who sacrifices their liberty deserves the support of people who can't go to the front-line. If you fly, here is an opportunity: Fly bad. Do good.

www.rebeloffset.com

www.ingramcontent.com/pod-product-compliance
Lightning Source LLC
Chambersburg PA
CBHW020322010526
44107CB00054B/1946